World of science

THE EARTH AND THE SOLAR SYSTEM

BAY BOOKS LONDON & SYDNEY

1980 Published by Bay Books
157–167 Bayswater Road, Rushcutters
Bay NSW 2011 Australia
© 1980 Bay Books
National Library of Australia
Card Number and ISBN 0 85835 268 0
Design: Sackville Design Group
Printed by Tien Wah Press, Singapore.

THE EARTH AND THE SOLAR SYSTEM

Our Earth is one of the planets of the *solar system,* which is made up of the sun in the centre, nine *planets* including the earth, and many smaller bodies. The smallest bodies that rotate around the sun are called *asteroids*, and fragments of these are *meteors.* Collections of dust and gas form *comets.*

All the objects in the solar system, including the earth, are under the gravitational pull of the sun which holds them orbiting, or turning, around it.

The formation of the solar system

There are several ideas or *theories* about how the solar system was formed. One theory is that the solar system was created at the same time as the sun, which is a star, when it began to form from a giant cloud of gas. This gas drew together to form a core around which the force of gravity collected an encircling disc of left-over gas and particles of dust.

The particles which made up this disc collided with each other and caused the formation of other small solid bodies. These attracted more material from the disc until there was enough to form the planets. The other objects in the solar system, such as the meteors, are the 'rubbish' left over from that time.

Our solar system is part of the *universe,* some of which we see when we look out at the night sky. No one knows just how big the universe is, but we believe it stretches farther than we can see with the biggest telescopes. In the universe there are many systems like our solar system.

This theory therefore suggests that planets are formed

The planets of the solar system extend from Mercury, nearest the sun, to distant Pluto. They are shown here in comparison of size: Earth is one of the smallest planets.

The Sun

The planets take different amounts of time to circle the sun according to their distance from it. Mercury orbits the sun $4\frac{1}{4}$ times a year but Pluto, more than 5000 million km away, takes about 360 years to make a single orbit.

as a result of star formation. In some cases, there are two stars in a system and, as far as we know, these have no planets, but there are plenty of single stars which we believe have planets rather like our own.

Galaxies and light years

A system of stars and planets like the one containing our solar system is called a *galaxy*. There are many galaxies in the universe.

There are great distances between the planets, and even greater distances between the stars. These distances are measured in *light years.* One light year is the distance light travels in one year. Since light travels about 300,000 km a second, a light year is a vast distance. The sun is eight light minutes away, which means the light from the sun takes eight minutes to reach us. Looking into space we are looking backward into time. We see the sun as it was eight minutes ago and the nearest galaxy to ours, Andromeda, as it was millions of years ago.

THE EARTH'S STRUCTURE AND CRUST

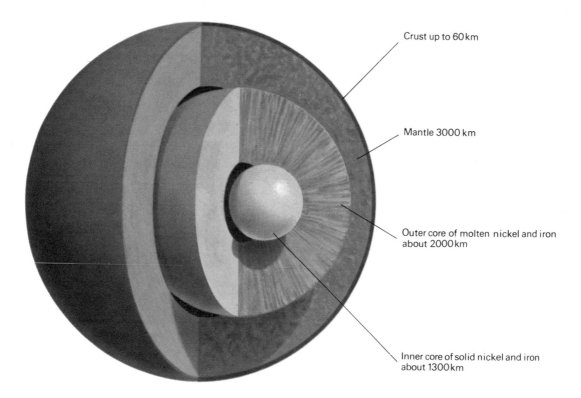

Crust up to 60 km

Mantle 3000 km

Outer core of molten nickel and iron about 2000 km

Inner core of solid nickel and iron about 1300 km

The earth's globe is made up of three layers. The surface layer is called the *crust*. Its thickness varies from about 8 km under the oceans to about 30 to 50 km under the continents. The next layer is the *mantle*, made up of harder, denser, rocks than the crust. It is about 3000 km thick. Next is the *core*, believed to be liquid outside, then, as far as we know, solid right at the centre.

If one could cut a section out of the Earth like cutting into a fruit, one would see the different layers shown in this diagram.

Igneous rocks

The rocks in the crust are classified into three main groups, according to how they were formed: *igneous* rocks, *metamorphic* rocks, and *sedimentary* rocks.

'Igneous' comes from the Latin word for fire, and igneous rocks are formed from molten material called *magma* that has cooled and hardened. Magma is molten rock. It probably comes from the mantle where solid rock

The strange hexagonal columns of the Giant's Causeway in Ireland are formed of basalt, a type of igneous rock created by volcanoes millions of years ago.

is turned into liquid through a combination of great heat and pressure. When this happens, some of the magma is forced up through cracks in the earth's crust. An active, or erupting, volcano forces out streams of magma in the form of *lava*. Lava cools very quickly when it reaches the air and becomes one form of igneous rock.

The type of igneous rock formed by cooling lava is called *extrusive*, because it is forced out or *extruded* from the interior of the earth. The speed at which the lava cools causes small crystals to form and the rocks are therefore smooth-surface or fine-textured like basalt and obsidian, although pumice stone is porous and frothy textured.

Another type of igneous rock is called *intrusive* and here the magma cools slowly beneath the earth's surface. This sometimes forms large masses called *batholiths*. These rocks are coarse-textured with large crystals resulting from slow cooling. Typical intrusive igneous rocks are granite, gabbro and syenite.

Minerals found in both extrusive and intrusive igneous rocks include felspar, hornblende, mica, quartz and pyroxene. Some of the oldest rocks on earth are igneous, but so are some of the newest. Every time you hear of a volcano erupting and pouring out lava, you know that when the lava cools new igneous rocks will be formed.

Sedimentary rocks

Sedimentary rocks are formed by the slow but never-ending process of *erosion* or *weathering* which breaks down the hardest rocks to small particles. Over millions of years these particles of rock are carried from their original site by the wind, rivers, glaciers and ice sheets to be deposited somewhere else. Eventually, layers of rock sediment are built up and become compacted and cemented together to form new rocks. Sedimentary rocks form a small part of the earth's crust, but they form a large part of the *surface* of the earth, because it is only on the surface that weathering and erosion take place. Sedimentary rocks are important because they sometimes contain *fossils* (the remains of plants and animals) which give clues to what life was like on earth hundreds of thousands of years ago.

Chalk is a sedimentary rock formed by layers upon layers of animal shells gradually settling on the sea bed. This is the Shakespeare Cliff, one of the famous White Cliffs of Dover which are made of chalk.

Metamorphic rocks

Metamorphism is a word that means change from one thing to another. *Metamorphic* rocks are rocks that were originally either igneous or sedimentary and have been changed by heat or pressure or a combination of both. Metamorphism is caused by the upwelling of molten magma, or by great movements of the earth that squeeze and crumple rocks into mountain ranges.

When magma forces itself up between cracks in the earth's crust, it is extremely hot and its heat bakes and hardens rocks around it. Sometimes the heat releases substances from the rocks and this also changes their composition. When mountains are formed, pressure crushes and grinds the rocks. When heat and pressure are combined, other changes take place. This is how we get marble, which was originally limestone; and quartzite, which was originally sandstone. Slate is another example, formed from compressed clay and shale. As slate is in layers which can be split easily, it can be used for blackboards and for roof tiles.

Marble is a familiar and often beautiful metamorphic rock, formed from limestone.

A volcano forces out lava **1** which hardens to igneous rock **2**. As it is eroded by wind **3** and water **4**, particles are washed to the sea **5**. These and other sediments settle and form sedimentary rocks **6**. Earth movements fold the rocks, **7**, and may heat or pressure them to form metamorphic rocks **8**.

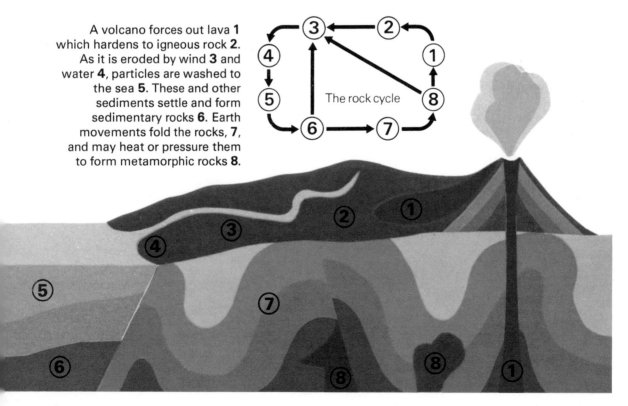

GEOLOGICAL TIME

The age of the earth

Until about 300 years ago, most people believed that the earth was formed quite recently. Some considered that the earth was created in 4004 BC, a date that was calculated from the number of generations of people listed in the Bible. Fossils found in some sedimentary rocks were thought to be of creatures that perished in the Flood that is described in the Bible.

In the 1880s, however, scientists began to realise that fossils were the remains of creatures which may have lived thousands or millions of years ago. They also came to understand that fossils represented the story of life on earth. But at that time no one could determine the true

Trilobites were crab-like animals that lived on Earth more than 200 million years ago. These fossils are the print of their shells, embedded in sediments that gradually formed into rock.

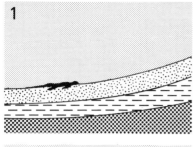
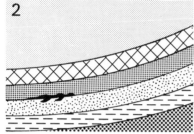
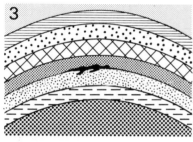
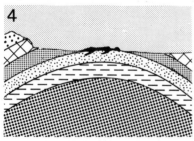

Formation of a fossil: when a creature dies its body falls to the sea bed **1**. There its shell is covered by layers of sediment **2**. These layers harden into rock which is shaped by Earth movements **3**. Eventually erosion brings the fossil back to the surface and our view **4**.

age of rocks and fossils. With the discovery of *radioactivity* and *radioisotopes* it became possible to fix the age of rocks by a process called *radiocarbon dating.* Now it is believed that the earth is about 4550 million years old and the oldest rocks about 3500 million years.

By studying rocks, scientists have been able to trace a broad history of the earth. They have identified periods of great volcanic activity, great changes of climate, periods when great mountains were formed and other periods when the mountains were worn down to flat plains.

Geology and palaeontology

Geology is the study of the earth, in particular the nature and distribution of the materials that form the crust. It has many branches. The study of *rocks* is called *petrology,* the study of minerals is *mineralogy.* There is also *geophysics,* which applies the laws of physics to the study of the earth, *geochemistry* which applies chemistry to the study of the earth's crust and *structural geology,* which studies the arrangement and structure of the rocks of the earth's surface. All of these branches of geology come under the heading of 'physical geology'. *Historical geology* covers the history of the earth, and includes *palaeontology,* the study of rock strata and fossils.

When scientists began the study of sedimentary rocks they discovered that, in some places, they are hundreds of metres deep. It takes a long time for sedimentary material to accumulate on the bottom of seas and lakes today, so this was the first hint that the earth is very, very old. Taking into account that when rocks remain undisturbed the oldest will lie at the bottom, the scientists realised that fossils could be used to establish the relative ages of the rocks. One of the pioneers in this work was William Smith, a British engineer. He discovered that, although some fossils occur in many *strata* or layers of rock, others, called *index* fossils, occur in only one layer. By finding index fossils in different rocks, geologists could establish that the rocks were of the same age.

Using Smith's methods, geologists began to classify rocks in order of age and were able to divide the history of the earth into six main *eras*. These eras were then divided into *periods*, and the periods were divided into *epochs.* These are named so that when they are discussed everyone, wherever they come from, knows which part of the earth's history is being talked about.

Geological eras

Rocks formed more than 570 million years ago are called *Pre-Cambrian* and they contain very few fossils. However, rocks formed in the seas of the *Cambrian* period, about 570 to 500 million years ago, are quite rich in fossils, proving that some forms of life were abundant at that time.

The word 'Cambrian' comes from *Cambria*, the old Roman word for Wales, because it was there that these rocks were first studied. In the Cambrian period, shallow seas spread over large areas, including most of Britain and about one-third of North America. In some parts of North America, movements of the earth's crust pushed and folded the rocks up into mountains such as the Appalachians, and the seas disappeared.

The age of the Earth is measured in eras, each of which lasted millions of years. During each era different forms of life evolved. Compared to other life forms, man is a newcomer to the planet.

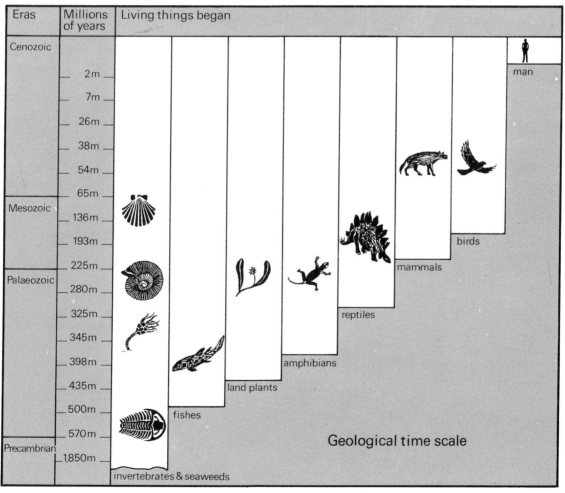

Geological time scale

The swampy forests that covered the Earth in the Carboniferous period must have looked like this.

Life at this time was of the *invertebrate* kind, that is, animals which have no backbones. Typical animals were the sea-dwelling brachiopods, molluscs, trilobites and an early type of coral. As far as we know there was no plant or animal life on land at this time.

The *Palaeozoic* era was a long period, lasting about 345 million years, during which plant and animal life began to evolve slowly on earth. The first *vertebrates* or **animals with backbones appeared.** These included fish, amphibians and reptiles. The first land plants that we know of also seem to have occurred during this period.

An important part of the Palaeozoic era is the *Carboniferous* period. It lasted about 65 million years and during this time the coal that we use today was formed from rich plant life that grew in swamps. Coal-forming plants included trees that grew to 30 metres tall, and giant ferns that are now extinct. Some big deposits of limestone were also formed during this period.

Beginning about 225 million years ago, the *Mesozoic* era lasted for 160 million years. For convenience, scientists divide it into three periods, the *Triassic,* the *Jurassic* and the *Cretaceous.* This was a very exciting evolutionary era, for it was then the dinosaurs made their appearance. The dinosaur is one of the largest creatures ever to walk the earth. Other spectacular creatures included the flying reptile called the pterodactyl, various marine reptiles and the first true bird. Primitive mammals also evolved during this time and flowering plants spread rapidly. In the seas were the spiral-shaped animals called ammonites, and coral.

The word 'cenozoic' comes from Greek words meaning 'new life', and the *Cenozoic* era covers the last 65 million years of the earth's history. Throughout this time most of the living things that exist today evolved. The great mountain ranges of the world, the Alps, the Himalayas and the Andes, were also formed during the Cenozoic era. It was towards the end of this era that man developed.

Life in the Cretaceous period included (left to right, below): Ichthyornis, an early bird; Pteranodon, a flying reptile; Triconodont, an early mammal; Triceratops, a dinosaur; ostrich dinosaur Ornithomimus; and Bennettitales, a primitive seed plant.

FOSSILS

What fossils are

Fossils are the evidence of life on earth many millions of years ago. They are always found in sedimentary rocks, which formed mainly in lakes, seas and swamps. Fossils are the remains of plants and animals, and in some cases almost the entire bodies of animals have been preserved.

Some fossils are *petrified* or turned to stone, some are *carbonised* or turned to carbon, and others are moulds or casts. Animal tracks and the marks made by waves and winds have also been fossilised in certain kinds of rock strata and formations.

The rocks containing fossils were once buried deep under other layers of sedimentary rock, but from time to time the earth's surface moves, such as when mountain ranges are formed, and the old rocks become exposed. Erosion wears away the rocks and we can see the fossils.

This fossil from the Dolomite Mountains of Italy is of a fish still found in tropical waters today. Through such clues one can trace how long certain creatures have existed on Earth.

Barnaby's Picture Library

How fossils are formed

Usually, when an animal dies in the open, wind and rain will soon crumble even the hardest bones. But if the dead creature is quickly covered over by sediment, it is protected from decay. When there are many fossils of sea creatures in rocks, it is usually because the sea bottom was covered in a thick, soft ooze. Dead creatures sank to the bottom into the ooze which protected and preserved them and later became sedimentary rock.

The early creatures of the Pre-Cambrian period were probably soft-bodied and so their remains were not so easily preserved. When creatures developed shells, bones and teeth, their remains could become fossilised much more easily.

Because animal bodies usually decay after death, few have survived in their original state. However, some woolly mammoths have been discovered in the frozen soil of Alaska and Siberia. Here the climate has been arctic for many centuries and the mammoths were preserved by the cold just as they died. In fact, even though the bodies were 30,000 years old when they were discovered, the meat was still fresh.

Fossils become petrified, that is, turned into stone when the dead creature is buried and the soft parts decay. Water containing a mineral such as silica seeps into the

Not only animal remains become fossils – this fossilised fern (above left) dates from the Carboniferous period. Even older is the fossil (above) of a lampshell, from the Devonian period 390 million years ago.

This petrified forest in Australia gives us an idea what the area was like millions of years ago. The plaster cast (inset) is of a mammoth which was found preserved in ice, in Russia.

spaces left by the decayed parts, like wax filling a mould. Over many years the silica completely fills the spaces and eventually becomes a stone replica of the animal. A similar process takes place with trees and plants. One famous example is the petrified forest in Arizona.

Looking for fossils

If you want to look for fossils, first find a place where sedimentary rocks are exposed, such as a canyon, a sea cliff or a quarry. A good way to start is to pick up and carefully examine small pieces of rock that have fallen down. If the fossil is in a big piece of rock you can chip the whole piece away with a chisel or a geologist's hammer, but the rock around the fossil itself must be removed very carefully with a smaller tool like an awl, for fossils are very fragile.

MINERALS AND FUELS

Minerals occur naturally in the rocks of the earth's surface. They may be either *elements* or *compounds.* An element is a single, pure substance. Gold is sometimes found as nuggets of the element. A compound is two or more substances bound together in such a way that chemical processes are necessary to separate them. Iron is usually found as a compound. Most rocks are mixtures of several minerals.

Native elements and ores

Arsenic, copper, iron, gold, silver and sulphur are all elements that may occur by themselves. Geologists call them *native elements.* Most minerals are compounds of several elements. Geologists have discovered about 2000 minerals, of which only about 100 are common.

Metals are minerals; and gold is a metal that does not rust. Because of this and its rarity it is used as the basis of most world economies. These are some of America's gold reserves.

Part of the process of making steel from iron and other metals. The liquid metal is poured into huge moulds where it cools enough to solidify, but at this stage it is still red hot.

Metals do not always occur in a pure, isolated form. For example, the metal lead must be extracted or refined from this mineral, called galena.

The two main groups of minerals are *metallic* and *non-metallic*. Metallic minerals contain native elements and *ores* from which minerals can be extracted. Examples of ores are haematite and magnetite from which iron can be extracted; bauxite (aluminium); galena (lead and silver); malachite (copper); pitchblende (uranium) and pyrites (iron and sulphur). Non-metallic minerals include graphite, gypsum, halite (rock salt), quartz, talc and diamond.

Identifying minerals

The best way to identify minerals is by chemical analysis in a laboratory, but geologists have devised some simple tests which can be carried out in the field to provide the first clues as to the presence of minerals. Laboratory tests

Bauxite is a mineral that looks like rock (see inset) but contains aluminium. It is mined (top) and the mineral processed to extract the metal.

Shiny haematite is a very hard mineral which is one of the world's main sources of iron.

are then carried out later on samples. Field tests to identify minerals use *colour* and *streak, lustre, hardness, crystal forms* and *cleavage*.

Colour can sometimes be misleading, but experienced geologists can often make a good guess based on colour. Azurite, for example, is one mineral that is always blue.

The *streak* of a mineral is the colour produced by scratching the surface to get a small amount of powder, which is often a different colour from the solid mineral.

Lustre is the way a mineral reflects light and with experience it is possible to identify, for example, diamonds, which have a brilliant lustre; quartz, which has a glassy or vitreous lustre; and some other minerals which have a metallic lustre.

Hardness is tested in the field by trying to scratch one mineral with another or with a metal intrument like a knife or file. Geologists use a *scale of hardness* called *Mohs' scale* to compare and measure hardness in the field. Mohs' scale lists ten common minerals, from talc, which is very soft, up to diamond, which is extremely hard. A mineral high on the scale will scratch all those below it but not the one above it, so it is possible to grade the hardness of mineral samples in this simple way.

Different minerals have different *crystals,* and large crystals are easy to identify because each mineral has its own distinctive form. Some minerals, like galena and halite, form a cube; some, like zircon, form twelve-sided crystals; some, like quartz, form long hexagonal crystals.

Cleavage is the way a mineral splits or flakes. Some minerals split easily, others do not. Cleavage runs parallel to the faces of the mineral crystal and is therefore not the same as fracture, or breaking, which may run in any direction in any mineral.

1	talc
2	gypsum
3	calcite
4	fluorite
5	apatite
6	orthoclase
7	quartz
8	topaz
9	corundum
10	diamond

Above: Moh's Scale of mineral hardness.

Below: minerals occur in many shapes and colours. These three crystalline forms are quartz (left), calcite (right, top) and albite (right, below).

Searching for minerals

Many geologists work as prospectors for mining companies, or governments. They search for water, metals, coal, natural gas, petroleum, uranium and other valuable minerals such as lead, zinc and copper.

The early prospectors lead lonely lives and suffered many hardships in their search for gold and other valuable minerals. Their simple equipment consisted of picks, shovels, and large, flat, sieve-like pans for 'washing' earth and minerals.

The modern prospector is usually a geologist trained in chemistry and physics. In the search for valuable metals, minerals, fuels and even water, he will use a wide range of scientific tools. Aerial photographs and pictures transmitted back from satellites are used to show up features of the *terrain*, or earth's surface, which cannot easily be seen from the ground. By studying the fossils, prospectors can tell the age of rocks; by detonating explosives in the ground and measuring the shock waves with instruments called *seismographs*, they can tell the density of the rock. For locating radioactive substances, such as uranium, they use *Geiger counters,* which respond to the amount of radioactivity in the area by clicking.

Modern prospecting methods are particularly important today in the search for more oil and one of the underground structures of particular interest is an *anticline*. This is an upfold caused by pressure in subterranean layers of rock. Layers of rock are pushed into shapes like arches or domes and caverns can form inside them. Natural gas and petroleum are often found under these domes.

In the rock formation called an anticline, rocks are forced up into arched shapes forming caverns beneath them, which often contain natural oil or gas Rigs (opposite) probe the rock to find such deposits.

oil   water

impermeable or non-porous rocks

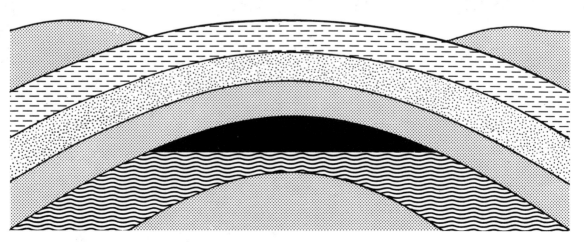

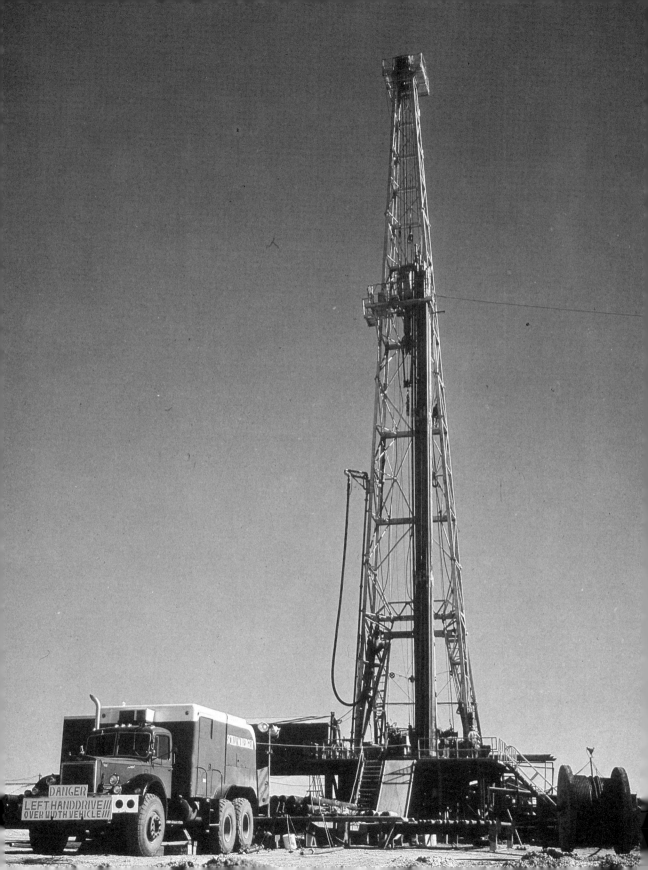

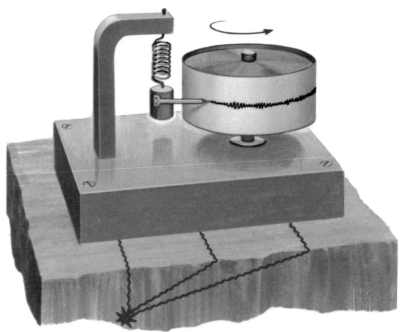

A seismograph (above) measures the strength of earthquakes. Seismographs round the world (below) can detect the location and strength of a 'quake. Seismographs are also used by prospectors for measuring the density of rocks.

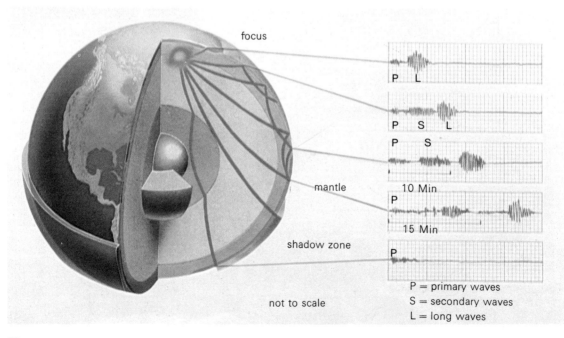

P = primary waves
S = secondary waves
L = long waves

Using minerals

The minerals in the earth are used for industry and commerce, to build houses, cities, factories, roads and machines, while natural fuels provide power for industry, transport and a host of other purposes.

Most metals are important to man, but iron, from which steel is made, has been the most important metal on earth for a very long time. Iron is usually found as iron ore, which has to be heated with coal and limestone in a furnace to make steel. This process is called *smelting*. Because it takes more coal than iron ore to make steel, the ore is usually brought to the coal, so countries rich in coal often import iron ore. Today, Australia is a country that is particularly rich in iron ore, far more than is needed in Australia alone, so it is exported to many countries, especially Japan, which has enormous steel industries. There are also huge deposits of iron ore in Russia and North America, and in a number of African countries.

Copper is another **very** important metal. It is essential to the manufacture of many electrical goods and machines because it is an excellent conductor of electricity. There are great deposits of copper ore in Africa and Canada.

Aluminium is also vital to modern industry. Light-weight construction is not only essential to aircraft and ships but also in many buildings, and the world uses huge amounts of aluminium. The ore of aluminium is *bauxite*, which is turned into the metal by the use of electric furnaces, not coal. So it is usually necessary to take bauxite to places where there is a good supply of cheap electricity, and this usually means hydro-electricity. Australia has huge deposits of bauxite such as those at Weipa on the Cape York peninsula. Much of it is taken to Bell Bay, Tasmania, where there is abundant hydro-electric power, for smelting it into metal.

Certain countries of the world have great resources of *fuels*. These are usually coal, oil and, in some countries, uranium for nuclear reactors. The main coal areas of the world for some hundreds of years have been Western Europe and North America.

Since the invention of the petrol engine and the motor-car and aeroplane, petroleum has been the world's most important fuel. It is a mineral oil found below the earth's surface. The first really big discoveries of petroleum were made in America. Oil production quickly became a huge industry there, but it was not long before oil was dis-

During the process of smelting, in which iron is combined with other metals to make steel, the metal is heated until it turns to a liquid and runs out of a hole at the bottom of the furnace.

Right: a marine oilrig. Above: a drilling crew at work. Their job requires great skill and may be dangerous.

covered in many other countries.

Today, the countries of the Middle East like Saudi Arabia and Iran are the greatest producers of oil. But the use of oil as a fuel has become so great that the great oilfields of Eastern Europe, North America and the Middle East are in real danger of running out of oil. This has caused a search for new oilfields all over the world, and many wells are now drilled below the sea. This is happening in the North Sea off the coast of Scotland, as well as in the Bass Strait which lies between the Australian mainland and Tasmania.

Sometimes coal is found on the surface of the earth, as here in Queensland, Australia, and can be mined by huge machines.

GEOMORPHOLOGY AND GEOPHYSICS

Geomorphology is the study of land forms and the natural forces that caused them. These forces can be *internal*, such as the effects of volcanoes and earthquakes, or *external,* such as the effects of erosion by wind, water, ice and other elements.

Ice ages and glaciers

Glaciers are moving bodies of ice. They shape the land by wearing away rocks, moving them and leaving them somewhere else. In North America and northern Europe, which now have temperate climates, great masses of ice moved across the land during the Ice Ages. The most recent Ice Age began about 600,000 years ago and lasted until about 15,000 years ago. Ice covered the land from the Arctic as far south as New York City in America and all of northern Europe, including all of the British Isles. An earlier Ice Age occurred in Pre-Cambrian times when most of Australia was covered with ice.

Glaciers tend to move downhill under the force of gravity. The pressure of the ice causes melting, and when the ice meets a rock water enters any cracks that may be present. When this water freezes again it expands and cracks the rock. When a glacier moves downhill, it also takes rocks with it. The glacier with its rocks becomes a kind of giant scraper, gouging out holes in the land and moving material in front of it like a bulldozer. In some parts of North America and Europe and Australia's Alps, there are hard rocks which have survived the Ice Ages but which show parallel scratch marks where glaciers passed over them. These marks are called *striations.*

When the ice melts away, great heaps of rock are often left. These are called *moraines.* Often they cause mountain lakes by damming streams.

Today, ice covers about one-tenth of the earth's land surface. If all the ice melted, the sea-level would rise by 60 to 90 m and most of the cities built on the sea, such as Melbourne, Sydney and New York, would disappear. In the Antarctic, the ice can measure from 300 to 1800 metres in depth.

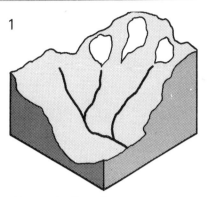

Above the line of permanent snow, pockets of ice are formed.

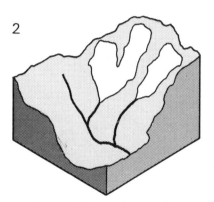

As they get bigger they push down the stream courses.

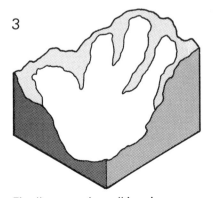

Finally several small ice rivers merge to form a glacier basin, slowly scouring down the valley.

Weathering

Still today, the earth's surface is gradually being worn away. In part of the weathering process, rain water combines with carbon dioxide to form a weak acid. The acid slowly removes some chemicals from certain rocks, causing them to weaken and, in some cases, dissolving them completely.

In hot, rocky deserts, rocks may shatter when their surfaces cool quickly in the evening. In moist, cool regions, rain water gets into the cracks and crevices of rocks and when there is a sudden freeze the water may turn to ice. When water freezes, it expands. Ice occupies more space than water, so the rocks are split apart.

This is the edge of a glacier, a slow-moving river of ice that grinds rocks smooth and carries boulders far from their source. Ice is a powerful agent of erosion.

In most places there is always water under the ground, called *ground water*, which seeps through underground layers of rock. Ground water often contains chemicals which will gradually dissolve some types of rocks over a long period of time. Limestone caves, some of them spectacular, are formed in this way.

Rivers are continually eroding the beds along which they flow. The mighty Mississippi River sweeps about two million tonnes of solid material past New Orleans every day.

Wind erosion is severe in the deserts and along some coastlines. Sand carried on the wind will, over a period of

River waters have cut down into the soft rocks to form Bryce Canyon, in the USA. The process of erosion has been continued by the wind, shaping the rock into weird pinnacles and arches.

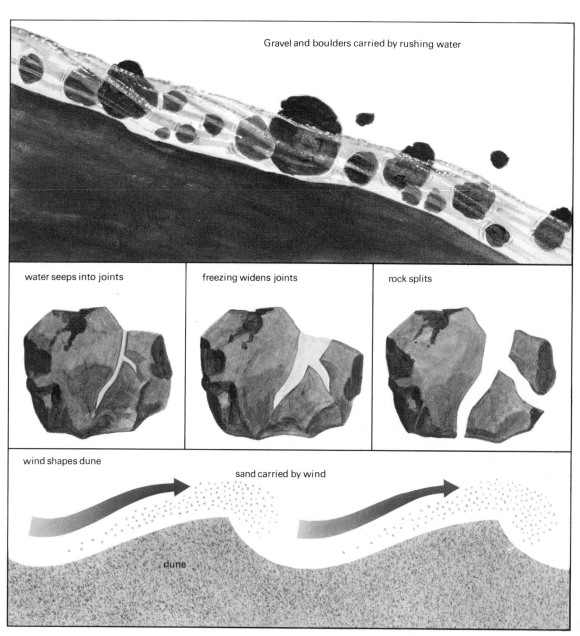

Processes of erosion: top, stones and boulders pushed by water grind against each other. Water may get into a crack in a rock (middle), expand as it freezes, and force the crack apart. Wind (below) blows sand to form dunes. As the sand is continually blown over the top, the dune moves forward. time, wear away rocks, especially soft sandstones such as those along the New South Wales coast. Coastlines are also affected by waves, tides and currents. At Surfer's Paradise on the Queensland coast, the beach is actually moving northward as the sand is carried out by the tides and deposited just a little further up the coast. If you walk around the rocks in almost any part of Australia's coastline, you will see where cliffs have fallen after a particularly severe storm accompanied by a high tide.

These stalactites have been created by the slow drip of water over hundreds of years.

Stalactites and stalagmites

Once a limestone cave has been formed, ground water usually continues to seep through the roof. Often this water contains dissolved limestone in the form of *calcium carbonate.* The water hangs from the roof in a droplet and exposure to the air causes some of it to evaporate. As a result a small quantity of calcium carbonate clings to the roof. The next droplet leaves a minute quantity of this solid material on top of the one before and gradually a long, tapering formation called a *stalactite* builds up. Droplets that fall to the floor of the cave also evaporate and leave deposits of calcium carbonate. These build upwards and are called *stalagmites.*

Volcanoes

Geologists define a volcano as a mountain that forms around a *vent,* or hole in the ground. These vents lead to the red-hot molten magma below the earth's crust. Some volcanoes are built around one central vent but most have several vents or one vent and several smaller *fissures* (cracks) through which molten rock, ash and steam are forced out.

Even today the Earth is changing. A volcano created the island of Surtsey (opposite) 15 years ago.

Some oddly-shaped mountains are created after a volcano dies. The central lava plug becomes a harder rock than the mountain around it, which was formed of volcanic ash. The soft rock is eroded by wind and rain until only the volcanic plug is left.

A volcanic eruption is one of the most spectacular natural events, but may cause tremendous damage and loss of life. On the other hand, some of the world's most fertile soils are formed from volcanic rocks and volcanic steam is used as a source of heat and power in such places as New Zealand, California, Iceland and Italy. A famous eruption was that of Mount Pelee in the island of Martinique in 1902.

Today there are about 450 active volcanoes on earth, some of which discharge large amounts of gases while others discharge great quantities of lava. Occasionally, huge amounts of rock are forced up by a newly formed volcano, probably caused by a crack in the earth's crust, and a new island or mountain may be formed. This happened in 1963, for example, off the coast of Iceland when a new island called Surtsey was formed. Some years earlier the new volcanic mountain of Paricutin arose in a farmer's field in central America. When a volcano dies, the once red-hot core cools down and becomes hard rock, usually much harder than the mountain around it. Over the years, the mountain may erode away while the hard core remains. These cores can be seen in such places as the Glasshouse Mountains in Queensland and the Warrumbungle Ranges in western New South Wales.

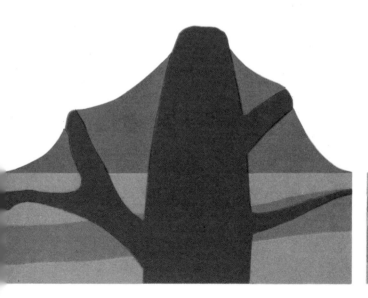
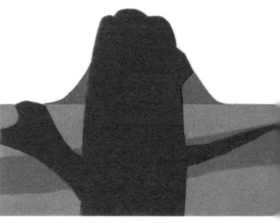

Earthquakes

Millions of years ago, earthquakes were more severe than they are today and many of our mountains were formed at that time.

Most earthquakes are associated with *faults* on or below the earth's surface in regions where the crust is unstable. A break or fracture may occur when there is movement of the rocks in the crust. The layers on the two sides of the fault move, either horizontally or vertically, or both, causing enormous upheavals on the surface. A movement of just a few centimetres vertically can cause a lot of damage, but some faults have moved as much as seven metres vertically and devasted whole cities. The city of San Francisco is built along a famous fault line called the San Andreas fault. In 1906 there was a huge earthquake that destroyed the city.

Geophysics and the shape of continents

Geophysics is the science that applies the laws and methods of physics to geology. One of the important branches of geophysics is the study of earthquakes and earth tremors, called *seismology*. Other aspects of geophysics include the study of the earth's magnetism, the effects of gravity, and meteorology (the study of weather).

Geophysicists attempting to understand how the surface of the earth was formed have produced the theory of *Continental Drift*. A look at a world map makes it easy to see how this theory first became popular. In fact, as early as 1620 the English scholar Francis Bacon noticed how the continents of the world seem to be like the pieces of a jigsaw puzzle. Since then, great similarities have been found in the fossils, rocks and mountain structures on both sides of the Atlantic, suggesting that at one time all the continents were joined together around the South Pole. According to this theory, the continents broke away from the Pole and drifted northward until they reached their present positions.

In more recent times a study of the coastlines of various continents has shown that the real edges of the continents are actually underwater, beyond the continental

The effects of an earthquake may be catastrophic, as in the town of Kakhk in Iran (opposite). Survivors comb the crumbled ruins in an attempt to rescue trapped victims and valuables.

shelf. By mapping the true edges of the continents, it was found that they fit together even better than appeared from the maps which show only the coastlines. Some scientists consider that India was once attached to Africa, but broke away and drifted, over millions of years, until it was forced against Asia. The force of the collision could have crumpled the surface and created the Himalaya Mountains.

It was also discovered that undersea mountain ridges were formed from new rock rising from below. Such a ridge runs north to south throughout the Atlantic Ocean and Iceland, which lies on the ridge, and is being widened by about 12 mm each year.

Many scientists now consider that the earth's crust consists of several rigid plates of rock about three to five km thick. These plates form the ocean floor and lie under the continents. Because there are movements in the earth's mantle below, these plates are being added to and moved sideways. When this happens, two plates can be pushed against each other and one has to move under or over the other, or if one is weaker, it will buckle.

This theory of moving plates could explain many of the earthquakes that occur around the world. The San Andreas fault in California could lie along the edge of two plates and the many earthquakes there may result from the plates moving against each other.

Scientists believe that our continents were formed as shown in the diagrams below.
1 At first, there was a great landmass in the middle of the sea. **2** Gradually the plates of the Earth's crust moved apart. **3** What we now know as Africa separated from South America, while **4** India floated up to collide with Asia, forming the Himalaya Mountains.

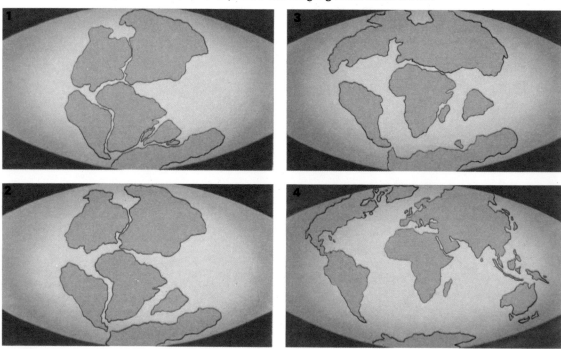

SEASONS, CLIMATE AND ECOLOGY

The zones of the earth

Most people on earth live in what are called the *temperate* regions. These are the parts of the earth's surface that lie about half-way between the Equator and the Poles. *Latitude* is the term used to describe the location between Poles and Equator, and much of Australia lies in the temperate latitudes between 20° and 40° south.

Nearer the equator and extending north and south of it are the hot *tropics*, and north and south of the temperate zones are the *polar* regions, the Arctic and the Antarctic.

Latitude and the seasons are two of the main reasons for the kind of weather we get in different places at different times. In the temperate zones we are accustomed to four distinct seasons, but in the tropics the differences are much less clear. People there usually speak of the 'wet' and the 'dry' season, because most rain comes at one time of the year when the winds called *monsoons* blow for several months. In the polar regions the change from season to season is so sudden that only the long winter and the short summer are recognisable.

The Earth's latitudes, the distance between the poles, are represented on a map of the world by parallel lines at different degrees from the centre of the globe (left). The middle parallel is known as the equator. Lines on the map radiating from pole to pole are known as meridians, and mark the degrees of longitude (below). Any place in the world can be identified by its degrees of latitude and longitude.

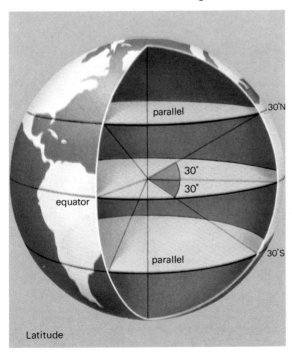

Latitude

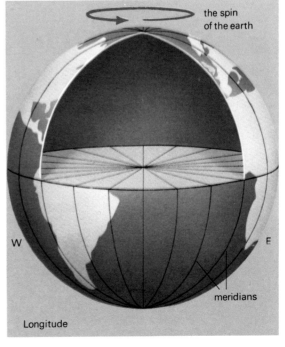

Longitude

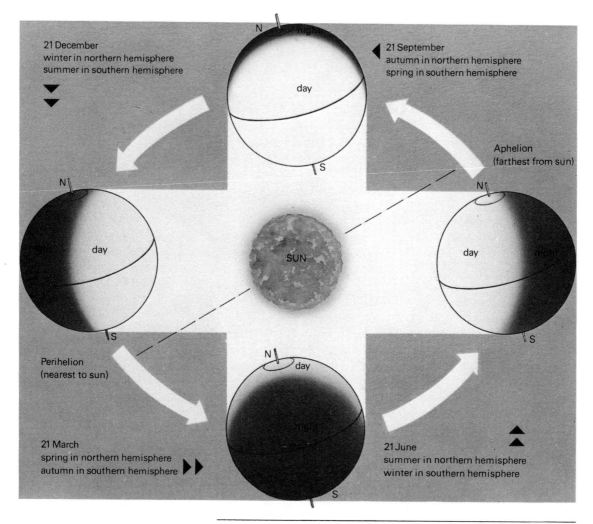

Seasons and weather

The seasons occur because of the way the earth rotates on its *axis* as it travels in its orbit around the sun. The axis is the imaginary line from Pole to Pole around which the earth rotates. It is tilted, not perpendicular. The angle of tilt, or *inclination* as it is called, is about $23\frac{1}{2}$ degrees. Because of this inclination, the northern hemisphere is sometimes pointed towards the sun, while at other times the southern hemisphere is pointed towards the sun. Thus at different times the different zones receive the sun's rays more directly or less directly. Also, the earth's orbit is not a true circle but a somewhat flattened circle, an *ellipse.* The seasons are also affected by the distance from the sun

Two other main influences on the weather are the winds and the size and shape of the continents. Generally,

As the Earth spins on its axis every 24 hours, we have daylight as long as our part of the planet faces the sun. Due to the inclination of the Earth's axis sometimes we are facing the sun more directly (which gives us summer) and at other times we are tilted away (our winter).

the largest continents have the greatest differences in temperature in the middle of their landmasses, and islands and sea coasts have the smallest changes in temperature. Thus in central Asia, North America and Europe there are cold winters and warm summers, while the climate on their coasts is less extreme. This is because the oceans gain and lose heat more slowly than the landmasses, and this helps to avoid great temperature changes on the coasts.

There are several great prevailing or regular winds on the earth's surface, and these too affect the weather. In the tropics, the south-west and north-east monsoons bring rain to great areas of India, South-east Asia and Japan, the islands of Indonesia, and parts of China and Australia. In other parts of the world, regular winds from the polar regions or the hot deserts of North Africa, for example, bring periods of cold or hot weather.

Ecology and environment

Millions of different kinds of organisms or living things live on the earth and each has its place in nature. But each animal, including man, plant, microscopic fungus and minute bacterium, affects other organisms and is affected by them.

The place where each organism lives is its *environment.* The environment influences the organism and the organism affects its environment. The study of all these effects and influences is called *ecology.* A group of organisms of different kinds living together in one kind of environment is called an *ecosystem.*

Some ecosystems are very small, but nevertheless are quite complicated. A pond, for example, can include many kinds of *micro-organisms,* as well as a number of higher forms of life such as the larvae of insects, molluscs including snails, and small swimmers like crayfish. The pond may be large enough for fish to live in and have water fowl on the surface.

Other ecosystems such as the forests and plains are very large. In the great grasslands of Africa there are, among the larger animals, *carnivores*, or meat-eaters, such as lions, and *herbivores*, or plant-eaters, such as antelopes. These seem to be the dominant or main living things, but there are many other small animals, as well as all plants, that are important to this ecosystem. The microscopic bacteria and fungi in the soil affect the fer-

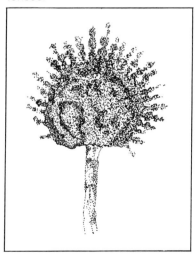

A microscopic fungus which is capable of eroding optical lenses.

In this diagram of a food web, arrows indicate what each creature feeds on. Gulls and gannets prey on the herring, which eats small fish such as the sand-eel and shrimp-like creatures such as hyperia. The sand-eel eats arrow worms, which eat barnacle larvae, which in turn eat the microscopic plankton life.

tility of the soil and the health of plants. When there is good plant growth, the plant-eaters have plenty of food and so the carnivores are also well fed.

There is always competition for space and food between the organisms in any ecosystem. In a small pond that is drying up in the summer, the environment at first favours many kinds of organisms. After a time the larger organisms, like the water fowl, frogs and many insects, find that the the available food supplies are getting short and many will leave the pond. Those who don't or can't leave, including the fish, will die from lack of food and sink to the bottom. Their bodies then provide food for the bacteria. However, many bacteria require oxygen, but this too becomes used up. Finally, only those bacteria that can live without oxygen are left.

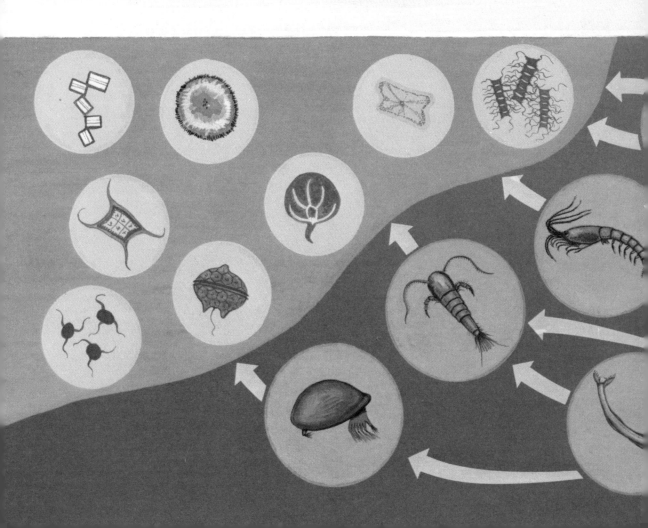

Colonisation

When an environment is changing it may favour more and more organisms. A new island formed by the outflow of magma from an undersea volcano usually begins as bare rock. The first creatures to find a home on the new island are usually small plant organisms such as lichens and blue-green algae. These can survive with nothing more than sunlight and minerals in the rock. Over many years the rock becomes weathered and some of it breaks down into soil. Seeds carried to the island by the wind and by birds can grow in the soil, producing plants which are food for larger animals which come to the island. Eventually the island can support a wide variety of species.

This process is called *colonisation*. Another example is the *repopulation* of forest areas that have been destroyed by severe bush fires. Some of the roots of the previous vegetation will have survived and will put up shoots. But all the ashes of the burned trees and shrubs may quite easily change the composition of the soil, perhaps making it more acid. Under these conditions, the environment may favour different plants from the original ones and so different plant-eating animals may come to the area.

These changing ecosystems are of great interest to people whose hobby is nature study, as well as to scientists. Interesting studies can be made in areas where the environment shows a wide, but fairly constant, variation. Near river estuaries, the water may change over a short distance from fresh to brackish to salt. These stretches of the river, including the river banks, will be homes for different populations of organisms, each suited to a greater or lesser amount of salt. Some kinds of animals and plants are adapted to large variations in the salt content of the environment and can be found all along the river.

Ecosystems are often easily upset, and populations of organisms can change. This, in turn, causes changes in neighbouring ecosystems, causing a kind of 'chain reaction'. No greater changer of the environment has appeared on earth than man. With our industries and the pollution they cause, we have altered nature profoundly and sometimes disastrously. In the last few hundred years, man has changed nature more than it has changed in the previous million years. With a greater understanding of ecology, however, we can perhaps undo some of the damage we have done.

A bush fire in Zambia (below) may change the ecosystem of the area. Trees and the creatures living in them will be destroyed. Some plants will grow again from shoots, but the soil, changed by wood ash and erosion, may be suited to new types of plants and therefore different types of wildlife.

INDEX

THE EARTH AND THE SOLAR SYSTEM 1-2
THE EARTH'S STRUCTURE AND CRUST 3-6
GEOLOGICAL TIME 7-11
FOSSILS 12-14
MINERALS AND FUELS 15-24
GEOMORPHOLOGY AND GEOPHYSICS 25-34
SEASONS, CLIMATE AND ECOLOGY 35-40

Page numbers in italics refer to a diagram on that page.
Bold type refers to a heading or sub-heading.

A

Active volcanoes 30
Alaska 13
Albite *19*
Alps 11
Aluminium 17, *17*, 23
America 15
Ammonites 11
Amphibians 10
Andes 11
Andromeda 2
Angle of inclination (of earth's axis) 36
Anticline 20, *20*
Appalachians 9
Arches 27
Ash 29
Asteroids 1
Axis, of earth 36

B

Bacon, Francis 33
Bacteria 37-38
Basalt 4, *4*
Bass Strait 24
Batholiths 4
Bauxite 17, *17*, 23
Bell Bay, Tas. 23
Bennettitales *11*
Bible 7
Birds 11
Blue-green algae 40
Brachipods 10
Bryce Canyon, U.S.A. *27*
Bushfire *40*

C

Calcite 19
Calcium carbonate 29
Cambrian period 9
Cape York 23
Carbon dioxide (acid forming) 26
Carboniferous period 10, 11
Carbonised fossils 12
Carnivores 37, 38
Casts and moulds (fossils) 12
Caverns 20
Cenozoic 3, 11
Chalk 5
Clay 6
Cleavage test (minerals) 18, 19
Climate **35**
Coal 11, 23, *24*
Coal-forming plants 11
Coastal erosion 28
Colonisation **40**
Colour test (minerals) 18
Comets 1
Compounds 1
Continental drift theory 33-34
Continents **33**, 33-34, 36, 37
  formation *34*
Copper 17, 23
Coral 10
Core, earth's 3
Crayfish 37
Cretaceous period 11, *11*
Crust, earth's 3, 34
Crust plates 34
Crystal 4, 19, *19*
Crystal forms test (minerals) 18

D

Devonian period 13
Diamond 17, 19
Dinosaurs 11
Dolomite Mountains *12*
Domes 20
Drilling 24

E

Earth 1
  age **7**, 7-8
  axis and orbit 36, *36*
  crust 3, 5, 34
  structure **3**, *3*
  zones **35**
Earthquake 22, 33, 34
Ecology **35**, 37, 40
Ecosystems 37, 38, *38*, 40, *40*
Elements 15
Ellipse 36
Environment **37**
Epochs 8
Equator 35, *35*
Eras of earth's history 8-9, *9*
Erosion 5, 6, 12, *27*, 28, *28*
Evolution 9

F

Faults 33
Feldspar 4
Fish 10
Fissures 29
Food web *38*
Fossils 5, 7, *7*, **12**, *12*, 13, 12-14, 20
  formation *8*, **13**
  petrified 13-14
Fracture 33
Fuels **5**, 23
Fungus 37, *37*

G

Gabbro 4
Galaxies **2**
Galena *16*, 17, 19
Geiger counters 20
Geochemistry 8
Geological eras **9**
Geological time 7
Geologist 20
Geology **8**
Geomorphology **25**
Geophysics 8, **25**, **33**, 34
Giant's Causeway 4
Glaciers **25**, *25*, *26*, 25-26
Glasshouse Mountains 30
Gold 15, *15*
Granite 4
Graphite 17
Ground water 27
Gypsum 17

H

Haematite 17-18, *18*
Halite (rock salt) 17, 19
Hardness test (minerals) 18-19, *19*
Hemispheres of earth 36
Herbivores 37
Himalayas 11, 34
Horneblende 4
Hydroelectricity 23

I

Ice 25
Ice Ages **25**

Icthyornis *11*
Igneous rock **3**, 4, *6*
 extrusive 4
 intrusive 4
Index fossils 8
Insects 37
Invertebrate 10
Iron 15, 17, 23
Islands 37

J
Jurassic 11

K
Kakhk, Iran *32*

L
Land plants 10
Latitude 35, *35*
Lava 4, 6
Lava plug 30
Lead *16*
Lead and silver 17
Lichens 40
Light years *2*
Limestone 6, 11, 29
 caves 27, 29
Longitude 35, *35*
Lustre test (minerals) 18-19

M
Magma 3-4, 6, 29
Magnetite 17
Malachite 17
Mammals 11
Mammoths 13, *14*
Man *9*
Mantle 3, 34
Marble 6, *6*
Mercury 2
Meridians 35
Mesozoic 11
Metal 15
Metamorphic rocks 3, **6**, *6*
Meteors 1
Mica 4
Microorganisms 37
Middle East oil 24
Mineralogy 8
Minerals 4, **15**, 17-20, *19*, 23
 identification **17**, 17-19
 metallic 17
 non-metallic 17
 searching for **20**
 use **23**
Mississippi River 27
Moh's scale of hardness 19, *19*
Molluscs 10, 37
Monsoons 35, 37
Moraines 25
Mount Pelee 30
Moving plates theory 34

N
Native elements **15**
Natural gas 20

O
Obsidian 4

Oil 23-24
Oil rig *24*
Ooze 13
Ores **15**, 17
Ornithomimus *11*

P
Palaeontology **8**
Palaeozoic era 10-11
Paricutin 30
Periods of earth's history 8, *8*
Petrified forest 12-14, *14*
Petroleum 20, 23
Petrology 8
Pinnacles 27
Pitchblende 17
Planets *1, 2,* 1-2
Plankton 38
Pluto 2
Polar regions 35, 37
Poles 33, 35
Pre-Cambrian 9, 13, 25
Pressure 6
Prospectors 20, 22
Pteranodon *11*
Pterodactyl 11
Pumice stone 4
Pyrites 17
Pyroxene 4

Q
Quartz 4, 17, 19, *19*
Quartzite 6

R
Radioactivity 8, 20
Radiocarbon dating 8
Radioisotopes 8
Repopulation 40
Reptiles 10
Rigs *21*
River estuaries 40
Rocks 3, 15

S
San Andreas Fault 33-34
Sand erosion 27
Sandstone 6
San Francisco 33
Seasons **35, 36,** *36,* 37
Sediment 5, 8
Sedimentary rocks 3, **5,** *6,* 8, 12, 13
Seismographs 20, *22*
Seismology 33
Shale 6
Siberia 13
Silica 13
Slate 6
Smelting 23, *23*
Smith, William 8
Snails 37
Solar System 1-2
 formation **1**
Stalactites **29,** *29*
Stalagmites **29**
Star 1-2
Steam 29

Steel *16, 23*
Strata (rock) 8
Streak test (minerals) 18
Striations 25
Structural geology 8
Sulphur 17
Surface of earth 5
Surtsey Island 30, *31*
Swamps 11
Syenite 4

T
Talc 17
Temperate regions 35
Temperature differences 37
Terrain 20
Theory of moving plates 34
Triassic 11
Triceratops *11*
Triconodont *11*
Trilobites *7*, 10
Tropics 35

U
Undersea mountain ridges 34
Universe 1
Uranium 17, 20, 23

V
Vent 29
Vertebrates 10
Volcano 4, 6, *6,* **29**, 30, *30,* 40

W
Warrumbungle Ranges 30
Weather 36
Weathering 5, **26,** 26-28
Weipa 23
Wet and dry seasons 35
White cliffs of Dover *5*
Wind erosion 27-28
Winds 36-37

Z
Zircon 19
Zones of the earth 35